DARK ENERGY EXPLAINED

THE MYSTERIES OF SPACE

DARK ENERGY EXPLAINED

GINA HAGLER

Enslow Publishing
101 W. 23rd Street
Suite 240
New York, NY 10011
USA
enslow.com

Published in 2019 by Enslow Publishing, LLC.
101 W. 23rd Street, Suite 240, New York, NY 10011

Library of Congress Cataloging-in-Publication Data

Names: Hagler, Gina, author.
Title: Dark energy explained / Gina Hagler.
Description: New York : Enslow Publishing, [2019] | Series: The mysteries of space | Audience: Grades 7 to 12. | Includes bibliographical references and index.
Identifiers: LCCN 2018019090| ISBN 9781978504547 (library bound) | ISBN 9781978505551 (pbk.)
Subjects: LCSH: Dark energy (Astronomy)—Juvenile literature. | Astrophysics—Juvenile literature. | Cosmology—Juvenile literature.
Classification: LCC QB791.3 .H34 2018 | DDC 523.01—dc23
LC record available at https://lccn.loc.gov/2018019090

Printed in the United States of America

To Our Readers: We have done our best to make sure all websites in this book were active and appropriate when we went to press. However, the author and the publisher have no control over and assume no liability for the material available on those websites or on any websites they may link to. Any comments or suggestions can be sent by email to customerservice@enslow.com.

Photos Credits: Cover GiroScience/Shutterstock.com; p. 5 d1sk/Shutterstock.com; pp. 6–7 Luke Peterson Photography/Moment/Getty Images; pp. 11, 13 Encyclopaedia Britannica /Universal Images Group/Getty Images; pp. 14, 19 Designua/Shutterstock.com; p. 16 Pictorial Parade/Archive Photos/Getty Images; p. 17 ullstein bild/Getty Images; pp. 22–23 Underwood Archives/Archive Photos/Getty Images; pp. 24–25 NASA/Hulton Archive/Getty Images; p. 27 NASA/Roger Ressmeyer/Corbis Documentary/VCG/Getty Images; p. 29 Mark Garlick /Science Source; p. 32 Bertil Ericson/AFP/Getty Images; pp. 34–35 NASA/JPL/CXC/SAO /MPIA/Calar Alto/O.Krause et al./Science Source; p. 42 Science Photo Library/Science Source; p. 45 NASA/Science Source/Getty Images; p. 48 NASA/JPLCaltech; p. 51 Bettmann/Getty Images; p. 53 banderlog/Shutterstock.com; pp. 56–57 Ted Thai/The LIFE Picture Collection /Getty Images; p. 61 Time Life Pictures/The LIFE Picture Collection/Getty Images; p. 63 Science & Society Picture Library/Getty Images; p. 67 European Space Agency/Planck Collaboration; pp. 68–69 Detlev van Ravenswaay/Science Source; back cover and interior pages sdecoret/Shutterstock.com (Earth's atmosphere from space), clearviewstock /Shutterstock.com (space and stars).

CONTENTS

INTRODUCTION

Cosmology is the study of the universe as a whole. It explains the way the universe started, the way it has behaved since, and the way it will behave in the future. Scientists, from physicists to astronomers, put their knowledge to work in this field. The result is theories such as the big bang and the possibility of multiple universes. These theories are important because they may give us an understanding of the very nature of reality.

Interest in the stars is not new. Humans have had an interest in the heavens since they first turned their attention to the night sky. Those early humans could not have known that they were on a planet that orbited a star. They could not have known that their solar system, comprised of that star (the sun) and the eight planets, their moons, and other matter that orbits the sun, was one of many solar systems. Neither could they have known that their galaxy—a system bound by gravitation—was one of many in the universe.

Today we know this and more. Our study of cosmology has shown that Earth orbits the sun, that the sun is just one of at least two hundred billion stars

The Latin name for the Milky Way is *Via Lactea*, "the road of milk." The Greeks called it *galaxias kyklos*, or "milky circle." They thought it looked like a milky patch of sky.

(any of which might prove to have a solar system of its own), that our solar system is part of the Milky Way galaxy, and that our galaxy is one of more than one hundred billion galaxies that make up our universe.

We are far more sophisticated than those first humans, yet we still have much to learn. It's not possible to test theories about the cosmos, so cosmology is a science of observation. Scientists can't recreate the big bang at a scale that matches the size of the universe. They can't test their assumptions about what will happen when a planet explodes by destroying a planet and collecting data. As a result, strong theories are vital to cosmology.

Isaac Newton (1642–1727) had his theory of gravitation. Albert Einstein (1849–1955) had two theories: the theory of special relativity and the theory of general relativity. Newton and Einstein both based their theories on the work done before them. Each of their theories gave a clearer explanation of the workings of the universe as observed until that time. Although each of their theories was sound based upon the knowledge of their times, each missed the fact that the universe is expanding and that the force of gravity alone is not holding everything in place.

As theorists are able to observe more and more of the universe, they question Einstein's theories. Do they explain everything? Might they need to use Einstein's cosmological constant? Could there be some sort of energy-fluid that fills space? Is it possible that Einstein was just plain wrong? None of these possibilities satisfied theorists, but they needed a working solution. They decided there might be something at work in the universe that they could not yet detect—something that caused differences they did not expect and still could not explain. They call this "something" dark energy.

Chapter One

Einstein, Noether, Hilbert, and Theoretical Cosmology

The quest for a theory that explains everything we observe about the motion of bodies and light in our universe has been underway for centuries. From the earliest time, people noticed the behavior of the planets in the sky and tried to make sense of the way things worked. In 1666, Newton announced his theory of universal gravitation to explain how planets move around the sun. As he explained it, the mass of the sun held the smaller planets in orbit around it. However, theorists working in the fields of electricity and magnetism in the late 1800s found that their observations did not fit the expected outcomes according to Newton.

Since his teens, Einstein had been interested in what it would be like to travel at the speed of light. Experiments into the speed and nature of light became more common in the late 1800s. These results did not match the results that scientists expected. The theories they had did not explain what they were seeing.

In 1675, Danish astronomer Ole Rømer (1644–1710) had shown that light has a finite speed (about 186,000 miles, or 300,000 kilometers, per second). In 1865, Scottish physicist James Clerk Maxwell (1831–1879) theorized that light moved in a wave. Einstein thought about what a universal speed limit would mean for physics. His first work, the theory of special relativity, explored the simplest case, that of a motionless observer observing an object moving in a straight line at an unvarying rate of speed.

It was soon after the publication of this theory in 1905 that Einstein published a second, shorter paper on the relationship between matter and energy. The algebraic equation that expresses this relationship is $E=mc^2$, or energy is equal to mass times the speed of light squared. Put another way, mass and energy are different forms of the same thing. Einstein knew his theory of special relativity applied to a case in which the observer was at rest and the object moved in a straight path at a speed not greater than the speed of light. He continued to work on a theory of general relativity that would cover all situations.[1]

Hermann Minkowski (1864–1909) was a former professor of Einstein's. In 1908, after studying Einstein's theory of special relativity, he presented the theory that space and time were two of the four parts of space-time fabric. The model he suggested was a four-dimensional model. Einstein at first said that Minkowski's four-dimensional work was "superfluous learnedness." He quickly changed his mind. Space-time (also known as tensor mathematics) was essential to Einstein's theory of general relativity.[2]

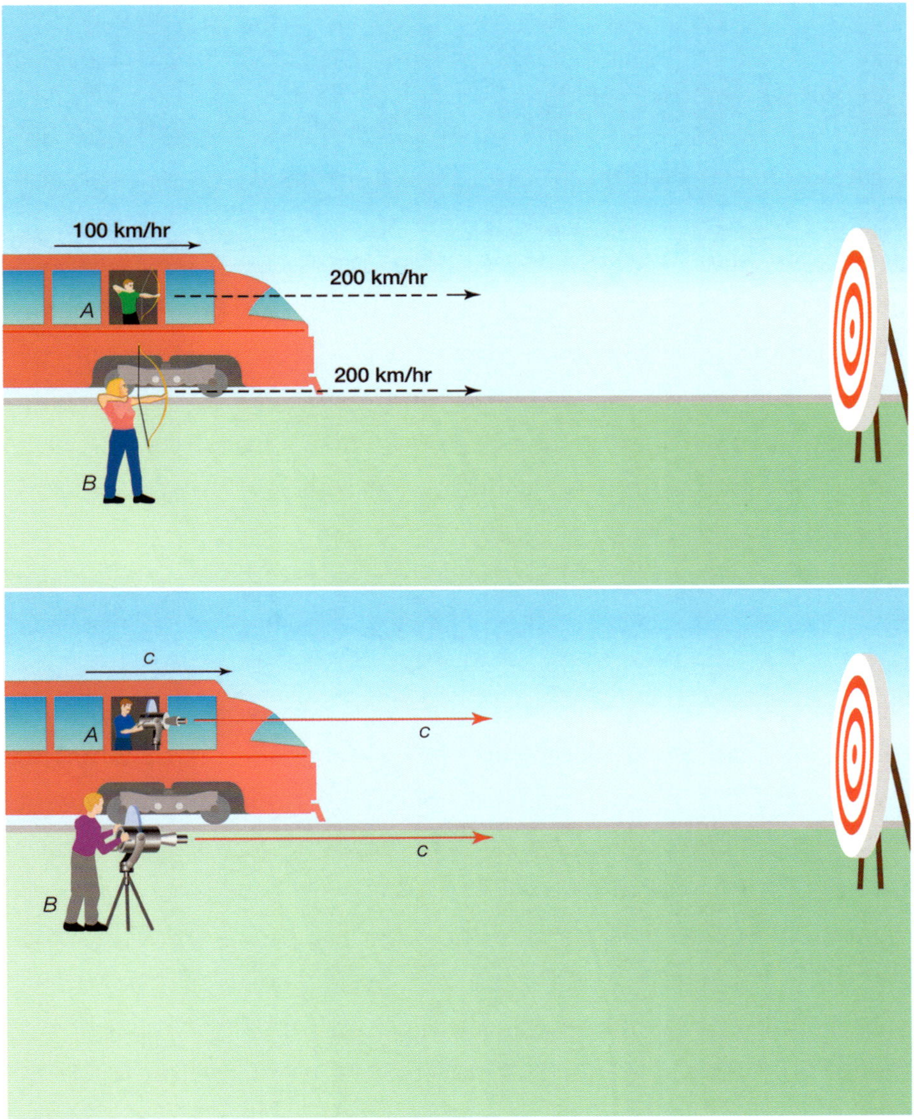

This illustration shows Einstein's theory of special relativity. In the top half of the image, the arrow shot from the train would arrive at the target first because the train's speed and the arrow's speed add together. The two beams of light on the bottom part of the image, however, would arrive at the target at the same time because the speed of light (c) is constant.

General Relativity

In 1907, while thinking it through, Einstein had what he reported as "the happiest thought of his life" while sitting in his chair at the patent office. He imagined a man falling from a height and realized that the man would not feel his weight as he fell. Einstein took this thought a step further. What if the man was in an elevator that was in free fall. The man inside the elevator would feel weightless. Anything else in the elevator with him would also float beside him. The man in the elevator would believe he was in a place that was free of gravity.

Einstein went even further. If the elevator had a rope attached to the top and that rope pulled upward, the man in the elevator would flatten against the floor. Any other objects in the chamber with him would also fall to the floor. The man would assume he was feeling the force of gravity yet be confused as to why the elevator did not hit the ground. When he noticed the rope at the top, he would deduce that he was in a stationary elevator that hung from a rope.

Einstein realized there was no way for the man inside the elevator to tell whether the force was acting on his inertial mass as he was being accelerated in a place without gravity or if his gravitational mass by being tugged downward by the force of gravity as the elevator hung suspended by the string. Since the mass in each case was equal, and there was no way to determine which was in effect, Einstein theorized that the man was in a gravitational field and that in such a field, the effects of acceleration and gravity were the same.[3]

Einstein's theory of general relativity was complicated. Its ten equations pushed the boundaries of algebra and geometry and required abstract algebra and non-Euclidian geometry, a type

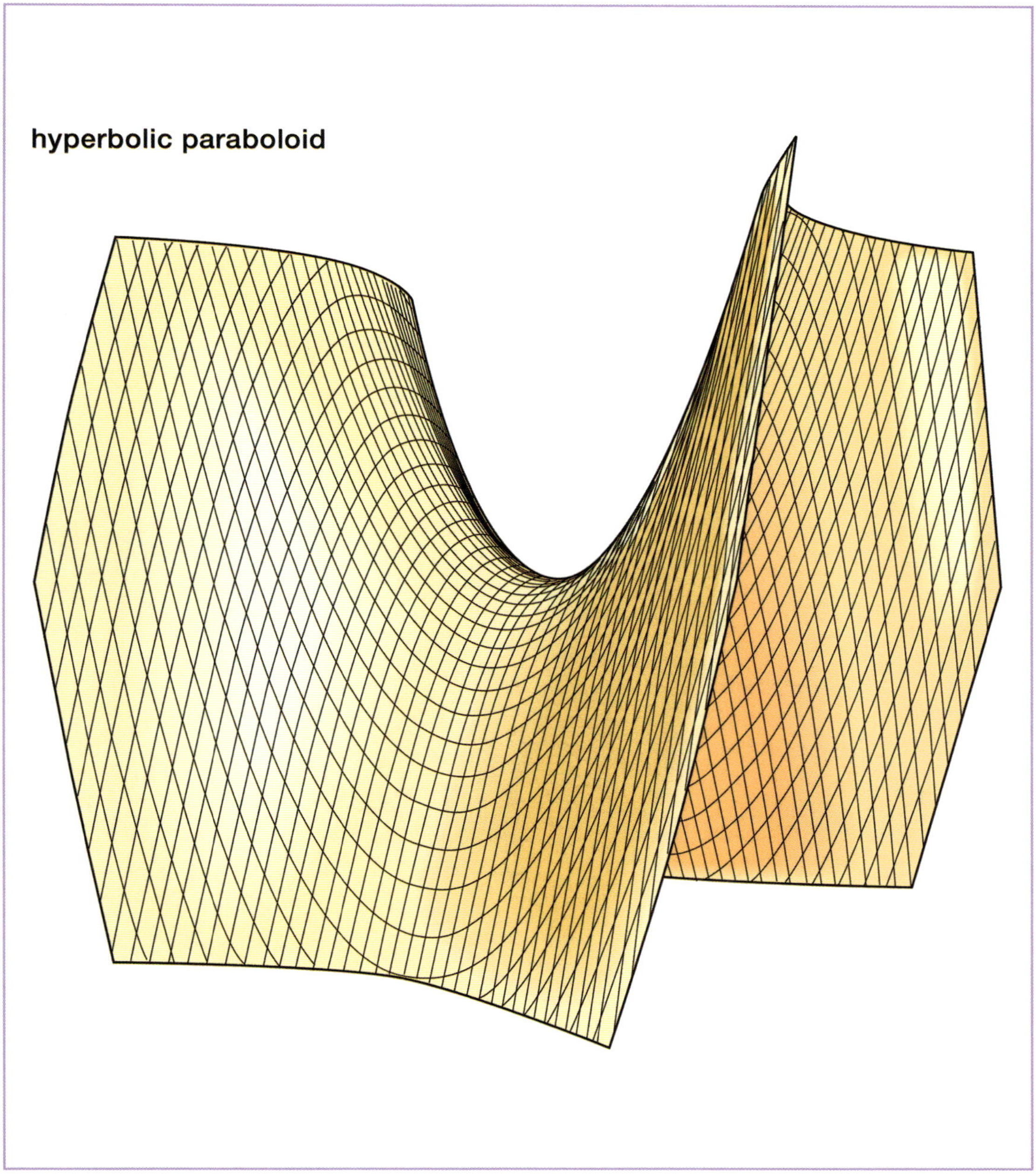

A hyperbolic paraboloid, which resembles a saddle, is an example of non-Euclidean geometry, a complex branch of geometry Einstein used for his theory of general relativity.

of geometry in which parallel lines either do not exist (spherical geometry) or in which there are at least two lines that pass through a point and are parallel (hyperbolic). The Einstein field equations

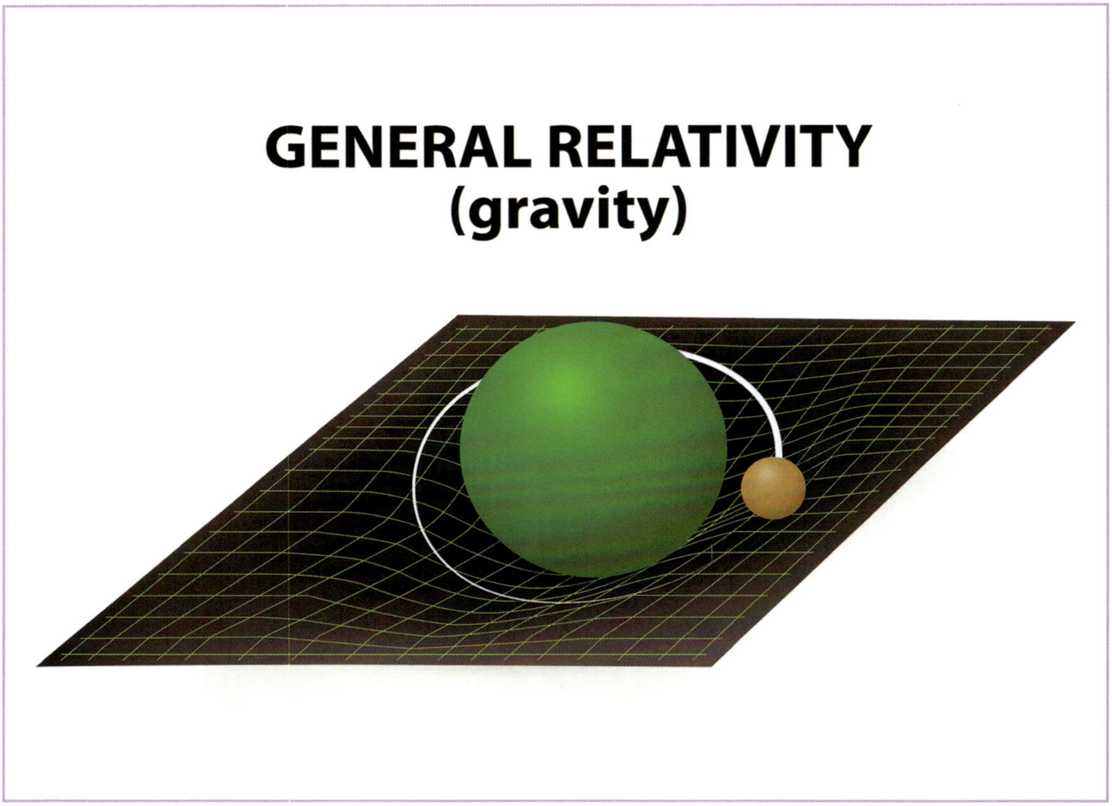

Einstein explained gravity as the curvature of space-time (the green grid in this illustration), while Newton had described it as a force. A small object moving through space-time would be affected by the curvature made by the larger object.

(EFF) detail the way mass interacts with space and time. Mathematicians such as Emmy Noether (1882–1935) and David Hilbert (1862–1943) applied their phenomenal mathematical minds to this work.

Einstein's theory of general relativity states that mass affects the shape of space and the flow of time. As a result, greater warping takes place with greater mass, in the same way a boulder will warp the surface of a trampoline more than a handful of pebbles. This theory of general relativity worked well, except for one thing. It was thought at that time that the universe was not expanding but rather in a stationary position, yet Einstein's calculations seemed to show that was not possible.

Einstein's Greatest Blunder

Einstein continued to struggle with the concept of an expanding universe. Nothing that had been observed up to that time indicated such a thing. In theorizing such a universe, Einstein was flying in the face of existing science. His own theory made it clear that space could never be at rest. Without a fudge factor, or a variable that could be changed to fit the situation, he could not balance the parts of his theory.

To solve the problem, Einstein used the cosmological constant lambda (Λ) as the fudge factor. At first, he set it to zero, but this resulted in an unstable universe. When it was set to a certain value, the universe represented was a stable universe. It continued to haunt Einstein that his equation needed this constant. It served simply to stand for whatever was causing the discrepancy in the theory without offering an understanding or explanation of the phenomenon.[4]

EMMY NOETHER AND DAVID HILBERT

Einstein was a gifted theoretician. He devised his theories by thinking problems through from start to finish and from every conceivable angle. For his theory of general relativity, groundbreaking work in mathematics was required to support his work.

Emmy Noether was a German mathematician whose work in abstract algebra was essential to the development of theoretical physics. Einstein knew and respected her, calling her the most "significant" and "creative" female mathematician of all time.[5]

Emmy Noether was one of the greatest mathematicians of all time. Noether's theorem united symmetry in nature and the universal laws of conservation.

David Hilbert was a German mathematician whom many consider among the greatest of his time. His work led to a new set of geometrical axioms (Hilbert's axioms). These postulates were vital to an understanding of the warping and folding of space-time.[6]

David Hilbert was equally renowned. His axioms replaced Euclid's axioms and brought two-dimensional and three-dimensional geometry into one system.

Friedmann's Fix

There were problems as others tested Einstein's theory of general relativity. In 1922, a Russian mathematician named Alexander Friedmann (1888–1925) proved that Einstein's universe must be unstable. His work revealed that any ripple or disturbance would force space-time to expand or contract. Because of this, the universe could never be static. It must always be in some state of change.

Einstein argued against Friedmann's findings before publicly stating that Friedmann's findings were correct. Friedmann worked within the general relativity framework to find that the energy of expansion in the universe is proportional to the sum of three types of energy: matter energy, cosmological energy, and curvature energy. Einstein's theory was the first to hint that the universe was expanding and had been expanding since the start. Friedmann's work agreed.[7]

By 1922, Edwin Hubble (1889–1953), an astronomer, measured the distances to other galaxies. Hubble was surprised to find that his measurements of distant stars proved that the universe was expanding. Hubble's findings caused a change in everything astronomers thought they knew. At the very least, it meant that the universe must have been smaller at one time and that Friedmann was correct. For the first time, it became clear that the universe had been expanding since the time of its origin.[8]

Learning that the universe was expanding caused Einstein to do away with his cosmological constant. He would later refer to it as the "biggest blunder of his life" according to physicist George Gamow (1904–1968). Whether the blunder was in creating the constant to begin with or in discarding what would eventually

The big bang theory states that the universe has been expanding since its highly dense and hot beginning. It is the result of the initial energy moving away from its point of origin.

prove to be the concept of dark energy is something no one will never know. What is known is that Einstein's error was not in his thinking, but in the limitations of knowledge at that time.[9]

Theoretical Cosmology

The groundbreaking work of Einstein, Friedmann, Hubble, and their colleagues opened the door to a new theory of the universe. This theory of the expanding universe led to the big bang theory, dark matter, dark energy, and other explanations of the origins and behavior of the universe. Professor Stephen Hawking (1942–2018) of the Centre for Theoretical Cosmology (CTC) at the University of Cambridge described the work done at the CTC in this way: "The CTC exists to develop theories of the Universe that are both mathematically consistent and observationally testable." This approach to cosmology is one that is shared by theorists around the globe.[10]

Chapter Two

Hubble and the Expanding Universe

Edwin Hubble lived at the start of the twentieth century. The work of this American astronomer changed what the world knew of the universe. It also provided proof of an expanding universe. During his lifetime, his work in astronomy did not make it possible for him to receive a Nobel Prize. However, his work was recognized, and Hubble's work later received awards from many respected scientific societies. His work continues to have an impact on our study of the cosmos to this day.

Contributions to Astronomy

Hubble is known for two significant findings that revolutionized astronomy. One had to do with the existence of multiple galaxies

in the universe. The other finding had to do with whether or not the universe was expanding. The result of both was that he was known as a modern-day Galileo.

When Hubble began his work in 1919, scientific teaching was that there was one galaxy—our galaxy. On December 30, 1924, Hubble shocked the world with his announcement that the Milky Way galaxy was one of many in the universe.

Hubble shocked the world again in 1929 when he published a paper with scientific proof that the universe is expanding. This observation lent support to Einstein's theory of relativity. In fact, Einstein traveled to Hubble's observatory to thank him personally.

Edwin Hubble, pictured here in 1947, used Cepheid variable stars to measure their distance from Earth. His work provided proof of an expanding universe.

Cepheid Variables

Both of Hubble's great discoveries were possible because of the powerful telescopes he had. These telescopes, which had nowhere near the power of those used today, made it possible to make observations of bodies and events in the night sky that had been impossible to observe up until that point.

Hubble focused his study on the faint stars in the Andromeda Nebula. Using Mount Wilson's 100-inch (254-centimeter) telescope, he could see that there were many stars there that were similar to the stars in our galaxy. The stars were dimmer because they were farther away. It was a specific type of star—the Cepheid variable—that made it possible for him to measure the distance.[1]

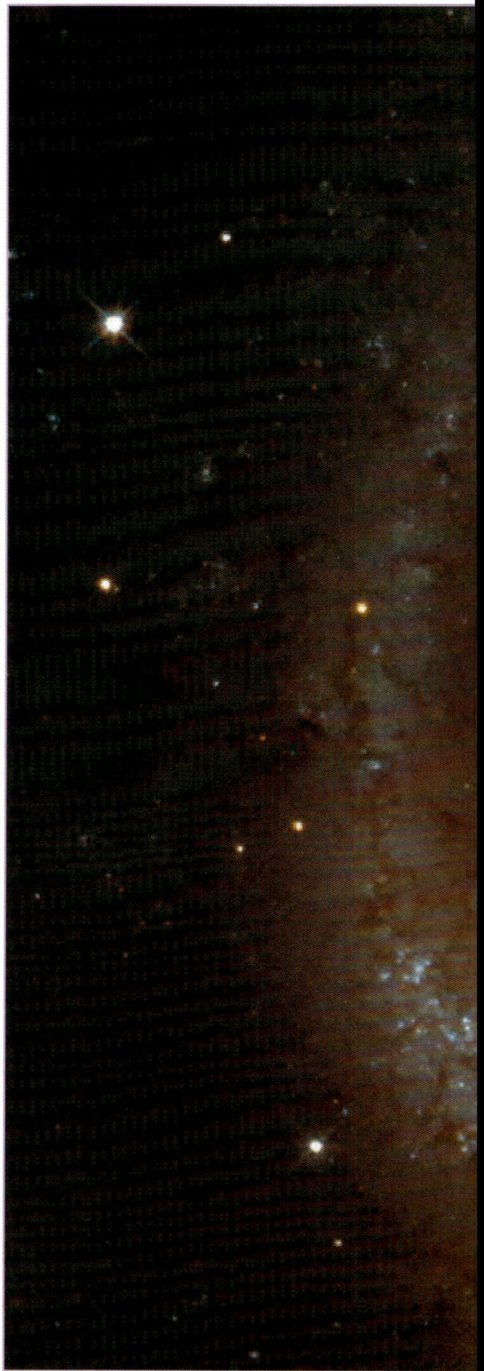

The Hubble Space Telescope captured this incredible image of galaxy NGC 4603, the most distant galaxy to contain Cepheid variable stars.

The Cepheid variables are stars that grow brighter and dimmer in a regular cycle. It was Henrietta Swan Leavitt (1868–1921), an American astronomer assigned to study the photographic plates of variable stars, who noticed in 1912 that the Cepheid stars that were the brightest had the longest period; they took the longest time to cycle from dim to bright and back again. It was she who discovered that Cepheid variables have the same degree of brightness, no matter where they are located. This property made these stars excellent for use in calculating distances.[2]

Cepheid variable stars can be observed from a great distance because they can be tens of thousands of times brighter than the sun. It is the expansion and cooling of the helium gas inside them that causes the change in their appearance. We know now that there are more than fourteen hundred Cepheids in the galaxies closest to us, as well as more in more distant galaxies.

THE HUBBLE SPACE TELESCOPE

The Hubble Space Telescope was launched in 1990. Upon launch, scientists discovered a flaw in the mirror. During a shuttle mission in 1993, that flaw was corrected. Ever since, it has sent back some of the most spectacular views of the universe ever captured. It is not the most powerful telescope ever launched, but it is the space telescope named for Edwin Hubble, who proved the existence of other galaxies, as well as the fact that the universe is expanding.

The scientific community also honored Hubble in 2011, when the Hubble Space Telescope was used to record observations of the V1 Cepheid that Hubble himself had used in his calculations. Any observations performed by the Hubble are expensive and take time away from other potential research opportunities. The fact that time was taken to capture these images is an indication of the respect felt for Hubble and his work to this day.

When the Hubble Space Telescope required repairs, the space shuttle *Endeavor* was used to reach it. Some of the most spectacular images ever seen were taken with the Hubble.

Hubble used the Cepheid variables to prove that the stars in Andromeda were 2.5 million light-years away. This distance put it far outside the boundaries of our own Milky Way galaxy, leading Hubble to understand that the stars in the Andromeda Nebula were actually in a galaxy of their own. He proved this by calculating the stars' intrinsic brightness and comparing it to what could be seen from where he observed. The difference in the luminosity he should have observed if a star was close by and the luminosity that was observed was caused by the distance to the star.

Hubble used one particular Cepheid variable, the V1, for his work. This specific star completes its cycle of pulsation every 31.4 days. By studying this star and applying the work done by Leavitt to his observations, Hubble proved there is more than one galaxy in the universe. Some had guessed there might be other galaxies, but most astronomers agreed there was only ours. One noted American astronomer of the time who believed in one galaxy, Harlow Shapley (1885–1972), wrote upon Hubble's announcement "Here is the letter that destroyed my universe."[3]

The Red Shift

Hubble also worked to study distant stars to observe their behavior. He noticed that some stars showed a red shift. This red shift is the result of a body moving away from us very quickly. The red comes from the stretching of the waves of light that are reaching us. The farther away the star is, the longer the waves have to be to reach us. This longer wave appears red to us on the spectrum of observable light. In the same way, the light from objects that are moving toward us appears blue. The waves are vibrating more quickly over the shorter distance.

To use these observations of red shift to determine the movement of objects in the sky, Hubble took what he discovered about the composition of the stars in question to calculate the part of the light spectrum he expected to observe. If the light visible to him were bluer than anticipated, that body would be moving toward him. If the light visible to him were redder than anticipated, that body would be moving away from him.

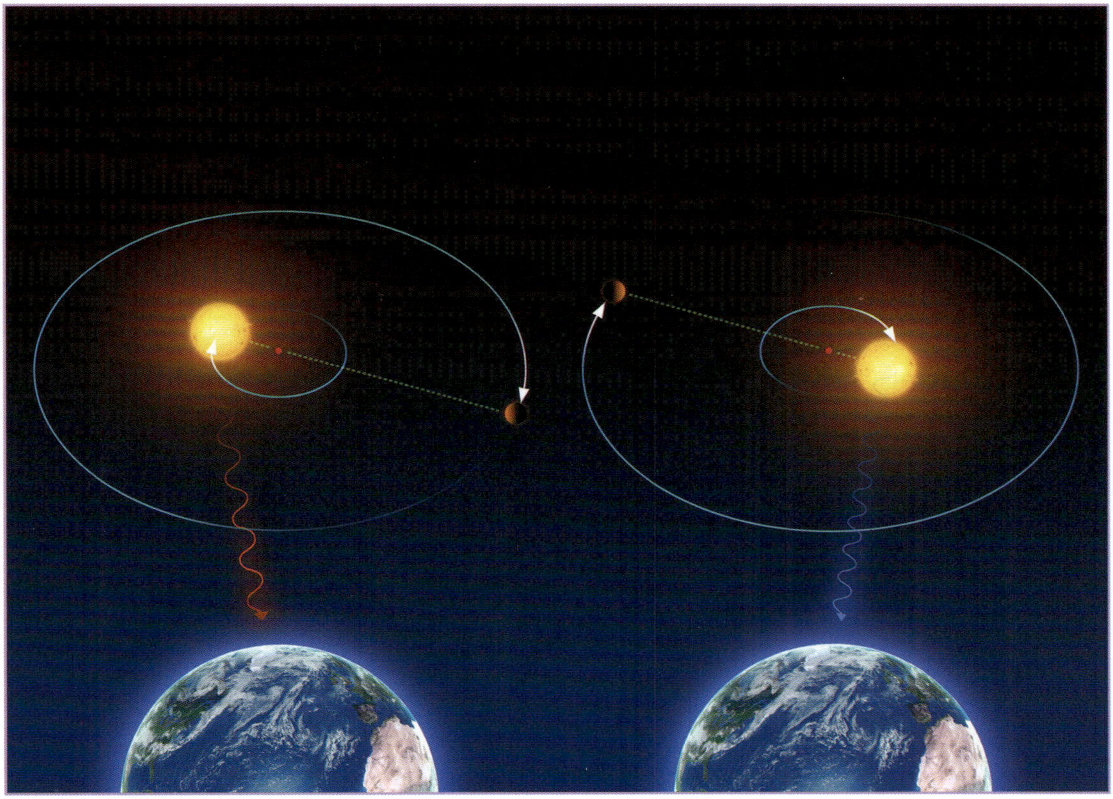

By observing the red shift, an observer can determine that a star is moving away from Earth. If the star is approaching Earth, the light is blue. This Doppler effect can be used to infer the existence of hidden objects.

Repeated observations proved time and again that bodies in the universe were moving away from us at a very high rate of speed. From these observations, Hubble was able to prove that the universe was expanding. This explained the problem with Einstein's earlier work—the way in which he could not get his theory to work without an extra variable. It turns out that variable was one that foretold the expanding universe.

The knowledge that our universe is expanding provided an opportunity for new ways of viewing the behavior of the universe, as well as what happened at the start of the universe. It led the way for theories such as the big bang, which states that our universe has been expanding since its instant of creation.

Hubble's Law

When Hubble calculated the rate of expansion in the universe, he used what is known today as Hubble's constant. This constant, calculated as 310 miles (500 km) per second per megaparsec, is thought to have been about ten times greater than the actual rate. (A megaparsec is about 3.3 million light-years.) It may be that a number that is more precise will be used in the future. Either way, Hubble's law is used to calculate the smooth expansion of the universe.

Hubble's proof of an expanding universe provided the logic behind the work of Newton and Einstein. The universe did not collapse because it had been expanding from the start.

Chapter Three

Supernova Searches and Accelerated Expansion

Hubble's work proved that the universe was expanding, but it did not provide the "why" of the expansion. That work would be done over the years, with the biggest breakthroughs occurring in the 1990s. These breakthroughs were the result of competing high-red-shift supernova search teams that searched for direct evidence of dark energy. Their findings honed our understanding of the universe while raising new questions.

The Race for Direct Evidence of Dark Energy

By the 1980s, the search was on for direct observation of the "stuff" that made up three-quarters of the universe. Scientists knew the something must exist. They just needed to find a way to prove it. They also needed a way to identify what that "stuff" might be.

From left to right, Adam Riess, Saul Perlmutter, and Brian Schmidt share the 2011 Nobel Prize in Physics for their work in dark energy using Type Ia supernovas.

During the 1980s and 1990s, one of the teams searching for dark energy was the Supernova Cosmology Project (SCP) led by American astrophysicists Saul Perlmutter (1959–) at UC Berkeley in California. This team planned to use Type Ia supernovas for their work.[1]

Another team searching for dark energy was the High-Z Supernova Search Team (High-Z SN Search Team) led by American-born Australian astrophysicist Brian Schmidt

(1967–) and American astrophysicist Adam Riess (1969–). This international group of twenty astronomers around the world made a collective effort to identify the source of the acceleration of the universe and learn more about the ultimate fate of the universe. They also chose to use Type Ia supernovas for their work.[2]

What Is a Supernova?

When a star dies, it doesn't fade away. It bursts into an explosion known as a supernova. These supernovas are incredibly powerful. They have more energy than the sun will have in its entire life. NASA refers to the explosion of a supernova as "the largest explosion that takes place in space."

Supernovas outshine everything around them. There have been supernovas reported from times before the telescope. Several dying stars go supernova every minute somewhere in our universe. When they explode, they deposit heavy elements such as iron, platinum, uranium, and gold in the universe.

The study of supernovas is an essential part of our understanding of our universe. These exploding stars may explain the source of the atoms in our bodies, the radiation that led to evolution on Earth, and the formation of our solar system.[3]

What Is a Type Ia Supernova?

The Type Ia supernova was chosen for the studies conducted by both teams. This type of supernova had properties that made it attractive to both groups. One of the reasons is that these are among the brightest things in the universe. Another reason is that they are very close to uniform in their brightness. As a

result, examination of the light from these cosmic explosions makes it possible for observers to calculate not only how far away they are, but how long ago they exploded.

This uniform behavior made them ideal candidates for standard candles, objects in the cosmos that give a known amount of light. The Type Ia supernova is ideal for this purpose because of the properties of the star involved. Astrophysicists know that this type of supernova always occurs in a star that is one of two stars orbiting each other. When one of the stars reaches a mass that is about 40 percent more massive than the sun, it explodes. The result is light that is five billion times brighter than the sun.[4]

Since the same thing occurs every time in the same way,

The Tycho supernova remnant, a Type Ia supernova in the constellation Cassiopeia, is all that exists from an explosion that occurred four centuries ago.

scientists can take the observed luminosity of the supernova and compare it to the known luminosity of the supernova. They then use these values to compute the distance to the supernova.

Using Type Ia Supernovas

In theory, using the light from a Type Ia supernova is simple. Observe the light, calculate the distances, and see what effect the new data has on the existing data. It was thought that, in practice, using a Type Ia supernova in this way was not quite like that because it would not be possible to catch the moment that was needed. It is also the case that supernovas of this kind are quite rare.

The SCP group found that they could spot stars about to go supernova by cycling through the same group of galaxies on a regular basis. As stars approached the moment of explosion, they grew lighter. After the supernova occurred, the light observed on Earth could be used in calculations as planned.

For its part, the High-Z SN Search Team was working to trace the universe's expansion over the past nine billion years. They made use of two telescopes to scan the sky every five minutes. They were looking for a level of faintness that would identify galaxies that might be home to the type of supernovas they sought.[5]

Ultimately, this team was also able to identify a sufficient quantity of these stars for their research needs. They were able to use their observations to calculate the rate of expansion over the nine billion years of the universe's existence. They, too, wanted to use the red shift of the light to show how the rate of expansion of the universe had slowed over time.

NOBEL PRIZE–WINNING WORK

In 2011, the Nobel Prize in Physics was shared by Saul Perlmutter, Brian Schmidt, and Adam Riess. The prize was "for the discovery of the accelerating expansion of the Universe through observations of distant supernovae."

Perlmutter was awarded half the prize, with Schmidt and Riess sharing the other half equally. These men were eligible for the Nobel Prize in Physics because, unlike in the time of Edwin Hubble, astronomy is now recognized as a field under the category of physics. As astrophysicists, their work could be considered and recognized.[6]

The red shift observed could be compared to the luminosity to calculate the rate of expansion of the universe—how much the universe has expanded since the star exploded. Both SCP and the High-Z SN Search Team expected to find evidence that the rate of expansion was decreasing. Neither expected the results would show that the expansion of the universe was not slowing down. Just the opposite. It was expanding at an increasing speed.

Not What They Expected

The results of their work with the red shift of the Type Ia supernovas did not show that the expansion of the universe was decreasing.

Deceleration had been expected because, as the thinking went, if the universe had been formed by a huge explosion, the rate of speed of the pieces flying apart should have reached a maximum and begun to slow.

Both teams saw that far from accelerating at a slower rate, the universe was accelerating at a growing rate. This could not be possible unless there was a source of energy that fell outside the types and sources of energy that were known at the time. Could Einstein's cosmological constant have been standing in for this other source of energy?

The findings of both teams matched, yet they had used different data and worked without knowledge of each other's findings. As mentioned on the High-Z SN Search Team site, these nearly identical conclusions could mean a number of things. It could be that both teams misinterpreted the supernova data in the same way. It could mean that both teams are right. It could be that the entire model used to explain the cosmos is incorrect.

These possibilities point to one of three possible conclusions: the universe is expanding at an accelerating rate, and the cause of this acceleration is the unknown energy known as dark energy; Einstein's theory of general relativity is wrong, causing everything based on it to be wrong; or there is something to do with the nature of supernovas that is misunderstood or has changed over time.

Other possibilities put forward by the SCP and other entities included the possibility that the universe has gone back and forth between expansion and contraction over time, that there is a dynamic form of energy in the universe with a changing value, or that Einstein's general theory of relativity is flawed.

For the scientists seeking answers about the nature of the universe and its implications, this state of uncertainty was highly unsatisfactory. More study was needed. More data was required. To truly understand and verify the data being recorded with the Type Ia supernovas would require a telescope that observed from outside Earth's atmosphere.

Such a telescope would not only operate outside the limitations of our atmosphere, it could observe supernovas that were closer than we could get while standing on Earth. The observations made would be of supernovas that were not so far into the red shift that it was difficult to "see" what the light was like before the acceleration. A telescope that met these requirements would provide the data necessary to test the science behind the initial findings.

With this in mind, physicists and astronomers from around the world worked together to design a satellite for the purpose of studying dark energy. This telescope, the SuperNova/ Acceleration Probe (SNAP), would be used to perform precise calculations on the expansion of the universe while studying dark energy and dark matter.[7]

Chapter Four

The Cosmological Constant

The cosmological constant, lambda (Λ), is the physical symbol used by the scientific community to represent unanswered questions about the expansion of the universe since its start nine billion years ago—specifically, unanswered questions regarding whether the universe is expanding, how quickly it is expanding, and whether or not any expansion has been constant over time. If the universe is expanding, the value of the constant is set one way. If it is not, the value of the constant is set another. The final answer to the value of this constant won't be in hand until astrophysicists have the data they require to determine the behavior of the universe over the past nine billion years.

Einstein's Lambda

Einstein introduced the cosmological constant in 1915, when he introduced his theory of general relativity. Also known as lambda (Λ), this fudge factor was set to a non-zero value so that his equations would represent a universe that was stable. This was necessary because Einstein's theory hinted at an unstable universe and the science of the time considered this to be impossible.

Einstein's theory of general relativity could explain, and even foresee, the behavior of gravity in terms of a space-time fabric. It could explain the effects of travel at the speed of light on the passage of time. It showed that Newton's concept of gravity was right—up to a point—but failed to take into account the way objects moved when near each other on the field that is formed by the "stuff" of the universe.

Einstein's theories of special and general relativity explained the relationships between acceleration and gravity (they are the same) and speed and time (each influences the effect of the other). He even proved, through $E=mc^2$, that energy and mass are two forms of the same thing.

Einstein was unsuccessful in only one area. He could not get his equations to work without the introduction of the cosmological constant. Without it, his theory proved that the universe was expanding. The only way to show that the universe was at rest was to use the cosmological constant as a fudge factor set to the value required to maintain that status.

Hubble Disproves Einstein

It wasn't long before Hubble came on the scene. By 1922, his work with Cepheid variable stars proved that the universe was expanding. This was as revolutionary to the field of astronomy as Galileo's proof that Copernicus had been correct and that the planets orbited the sun. Nothing up until then had proven this as

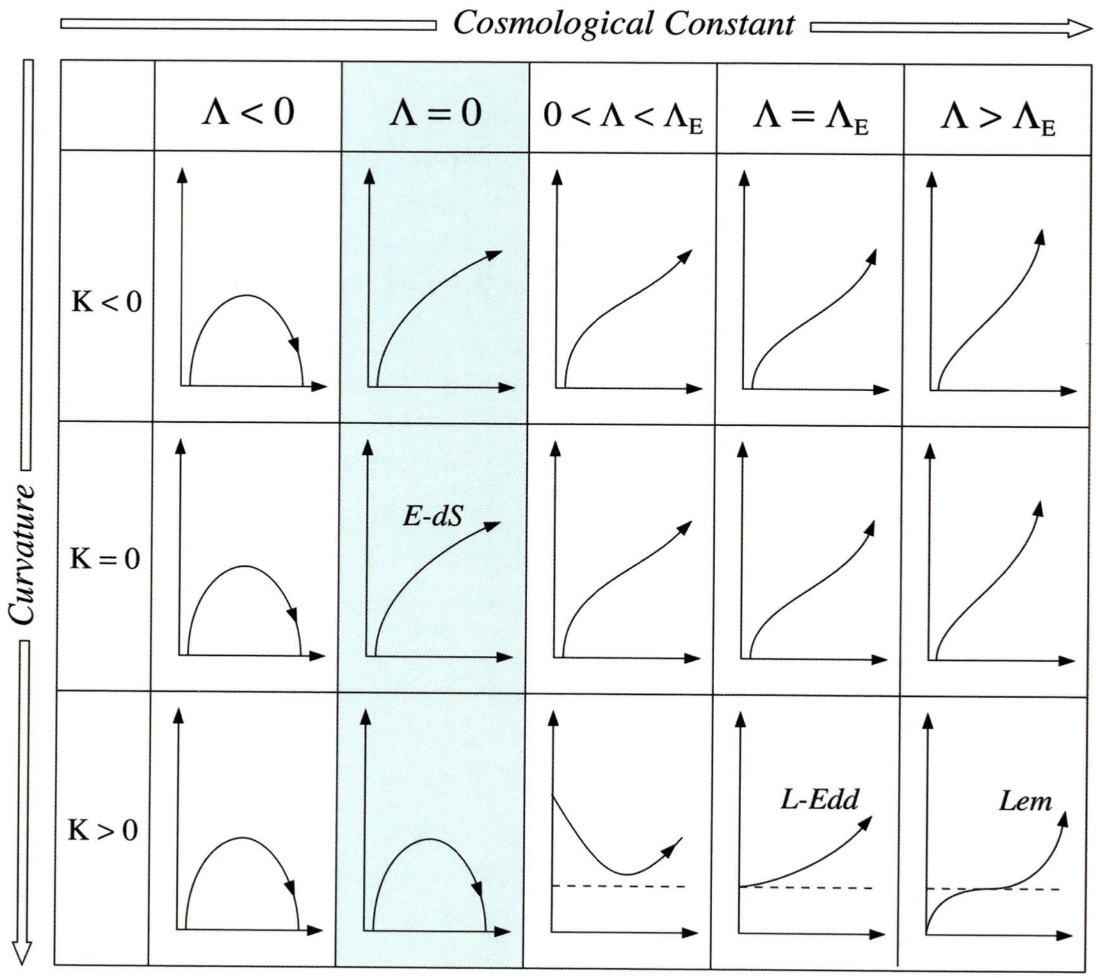

This chart illustrates the variation in the properties of the universe that are brought about by variations in the cosmological constant, known as lambda (Λ). The blue column has lambda set to zero.

a possibility, although Einstein's theory of general relativity had hinted at it.

Hubble's findings meant that Einstein's cosmological constant must be set to zero. By doing this with the fudge factor, the effects of this other force could be expressed, and the mathematical models showed that the universe was expanding.

As a result of Hubble's findings, Einstein got rid of his cosmological constant, since it would be set to zero and he no longer required a fudge factor.

WHAT IS A FUDGE FACTOR?

Have you ever had a math problem that just wouldn't work? You tried it different ways. You adjusted the values. You did everything you could think of, yet the equality you were trying to solve just wouldn't balance. Instead of having two numbers that were the same on either side of the equal sign, you had two numbers that were different no matter what you did?

This was Einstein's problem. He had his series of equations. He had his theories and other parts of the equations that he could prove through logic and through math. This was important because Einstein often used "thought problems" when creating a theory. He used pictures and situations in his head to envision how things worked. If his equation was off, there was a problem in his thinking.

(continued on the next page)

(continued from the previous page)

To get around that, Einstein used a fudge factor, a number that could be plugged in to make the equation work. The number didn't reflect any theory or explain anything about why the fudge factor was needed; it just made the equation work. Einstein called this fudge factor his cosmological constant.

Type Ia Supernova Teams

The work of the American Supernova Cosmology Team (SCP) and the international High-Z Supernova Search Team (High-Z SN Search Team) resulted in findings that support the concept of a universe that is expanding at an accelerating rate. The teams were not working together. They shared neither data nor ideas, yet each team, using the same type of supernova—the Type Ia supernova—arrived at the same conclusion.

Neither team could point to a specific force or type of energy or matter that could cause this accelerating expansion. Both teams agreed there had to be something, probably in the 75 percent of unidentified "stuff" in the universe, that accounted for this. They attributed it to the dark energy in the universe.

The Expanding Universe

The use of a non-zero cosmological constant also made it possible to model a universe in which galaxies and galaxy clusters are spread over a wide area. There had to be some way

Galaxy clusters are spread over a wide area in this image from the Hubble Advanced Camera for Surveys. The background galaxies (blue arcs) have had their light bent and amplified by the cluster's strong gravity.

to mathematically express the reality, that all of the objects in the universe were not in neat and concentrated proximity to each other.

The non-zero value of the cosmological constant allows the representation of these relationships by accounting for the

WHAT IS THE VALUE OF EINSTEIN'S COSMOLOGICAL CONSTANT?

In mathematics, a constant is a number that represents a relationship that does not vary. When used in a calculation, it has a specific meaning.

Some of the constants you may know include π. It is equal to the ratio of the length of a circle's circumference to its diameter. The value of π is 3.14. It does not change with different circumstances. It is always the same. Another constant you may know from geometry is the golden ratio, φ. It results in a numerical value of approximately 1.62.

Einstein's cosmological constant is the ratio of the actual density of the universe to the critical density, the value at which the universe stops expanding. The cosmological constant is estimated to be about 0.7.[1]

existence of an as-yet undefined entity that helps to explain the entire relationship.

It would be nice if the specifics of that entity were known and could be identified and studied further. Until that is possible, the use of the cosmological constant allows astrophysicists to go on with their work on the expanding universe. The cosmological constant serves as a marker, an indication that something is not explained fully and may not prove to be what scientists expect, but that it is exerting some influence or having some effect upon the way the universe behaves.

Dark Energy

Today the cosmological constant is seen as a representation of the dark energy in the universe. This dark energy is theorized to comprise 75 percent of the "stuff" of the universe. It is thought to be what causes the universe to expand, although the precise way this occurs is not yet understood.[2]

There are a variety of theories about dark energy. These theories range from miscalculations in current theories due to incomplete understanding of the objects studied, to the possibility of another universe that is influencing our own. Intriguing as they may be, none of the theories put forth can be proven until better data about the movement of objects in the cosmos are available.

Meanwhile, as scientists are better able to observe the activity of bodies near to us from outside Earth's atmosphere, the findings of those observations can be used to verify and improve upon current findings. This work may even help to frame new questions that will bring us closer to a complete understanding of our origins and future.

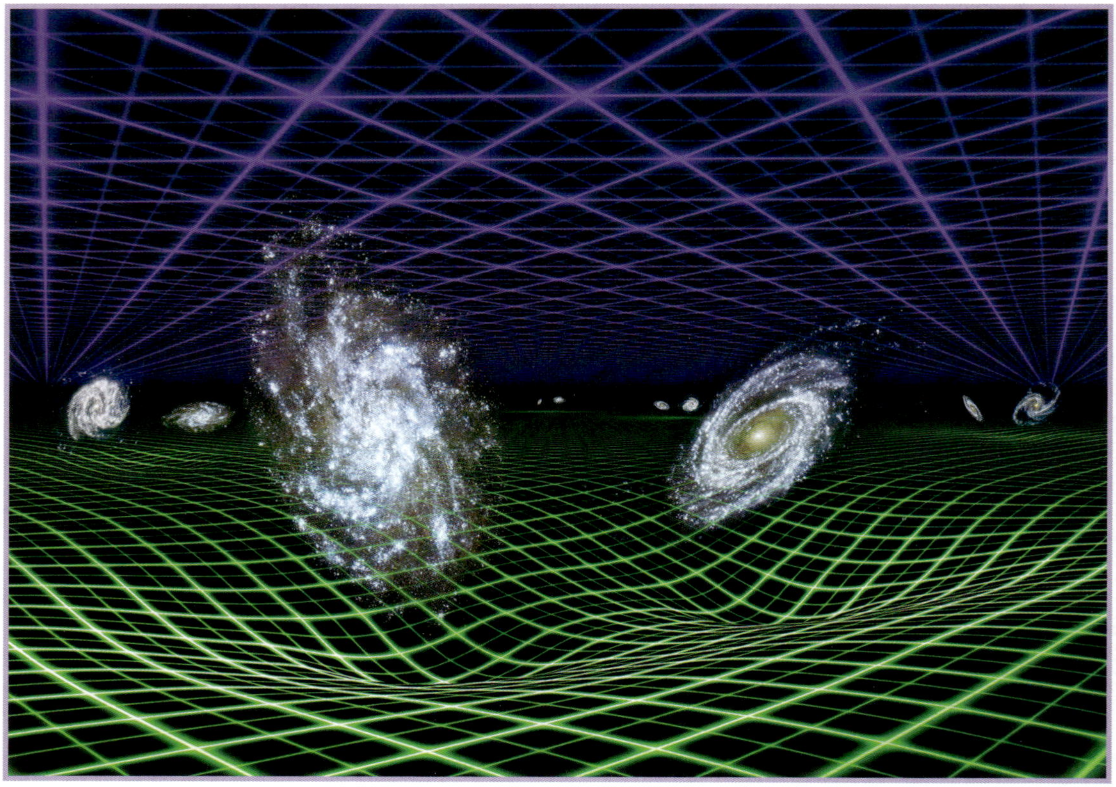

In this image, the purple grid represents the dark energy that pushes the cosmos apart. The green grid represents gravity in space-time. The galaxies in the middle are subject to the effects of both.

Through it all, the cosmological constant, lambda (Λ), has served as one piece in the solution to the problem of an unsteady universe. It was the expression that was required when knowledge of the universe at the start of the twentieth century was insufficient to accurately model the status of the universe at that time. It is the expression that is set to zero in our present time, when the status of the universe is thought to be one of expansion. It may prove to have other values as our knowledge of the universe grows.

Whatever its final value or use, the cosmological constant will remain a symbol of the uncertainty that is part of our attempts to understand the behavior of the universe. Since Einstein first put his theory forward, there has been a need to identify the unknown quantity in our understanding. Indeed, there is always a need to identify that which is not known in any scientific process. In this case, the unknown represented by the cosmological constant is the information that is key to our understanding of the expansion of the universe.

Chapter Five

Evidence of the Big Bang

For centuries, observers have wondered about how the universe was created. With the introduction of Einstein's theory of general relativity and the use of the cosmological constant, the door to an expanding universe was opened. Once Hubble proved there were not only multiple galaxies, but that they were moving apart, the question of how this all began consumed astronomers, physicists, and others with an interest in the new field of cosmology.

The Start of a Theory

After Einstein's theory of general relativity was released and before the results of Hubble's work with the Mount Wilson Observatory telescope were known, Belgian astronomer Georges Lemaître (1894–1966) proposed a theory of his own. This Catholic priest and physics professor suggested that the world was expanding.

His suggestion was that the universe was not only expanding but was expanding from a singular point. He called this the

George Lemaître, pictured here in 1933 in Washington, DC, did the work that formed the basis for the big bang theory. It was his effort that brought together the work done by Einstein and Hubble.

"primeval atom." He theorized that if the universe was expanding now, it must have been smaller or more compact in the distant past. Ultimately, something caused this compact point of origin to explode.

Lemaître's work formed the basis of what is known as the big bang theory. The name itself was introduced after Lemaître's ideas of the expanding universe—first hinted at by Einstein, then proven by Hubble—was referred to in the 1950s. The term was used by British astrophysicist Sir Fred Hoyle (1915–2001) during one of his radio lectures. He did not believe in the theory and used the term to show how ridiculous he found it.[1]

The big bang theory became the accepted theory of the origin of the universe. It uses Einstein's uncertainty about his equations, along with Hubble's discoveries, to trace back to what was then thought to be the inevitable conclusion that it all not only began with a big bang, but from one central point.

THE BIG BANG THEORY

The big bang theory is an accepted explanation for events that occurred at the birth of the universe. According to this theory, about 13.8 billion years ago, there was a single point of hot, dense mass. This point was very small—just a few millimeters in diameter. This mass, made up of all the matter and radiation that would be seen in our universe, existed for a fraction of a second before it exploded.

This bang was the start of everything we know. Within a trillion-trillionth of a second, the universe expanded to a huge scale. This expansion continues to this day. As the universe cooled, nano matter, then atoms, then stars and galaxies formed. The light from the big bang is still present in cosmic microwave radiation.[2]

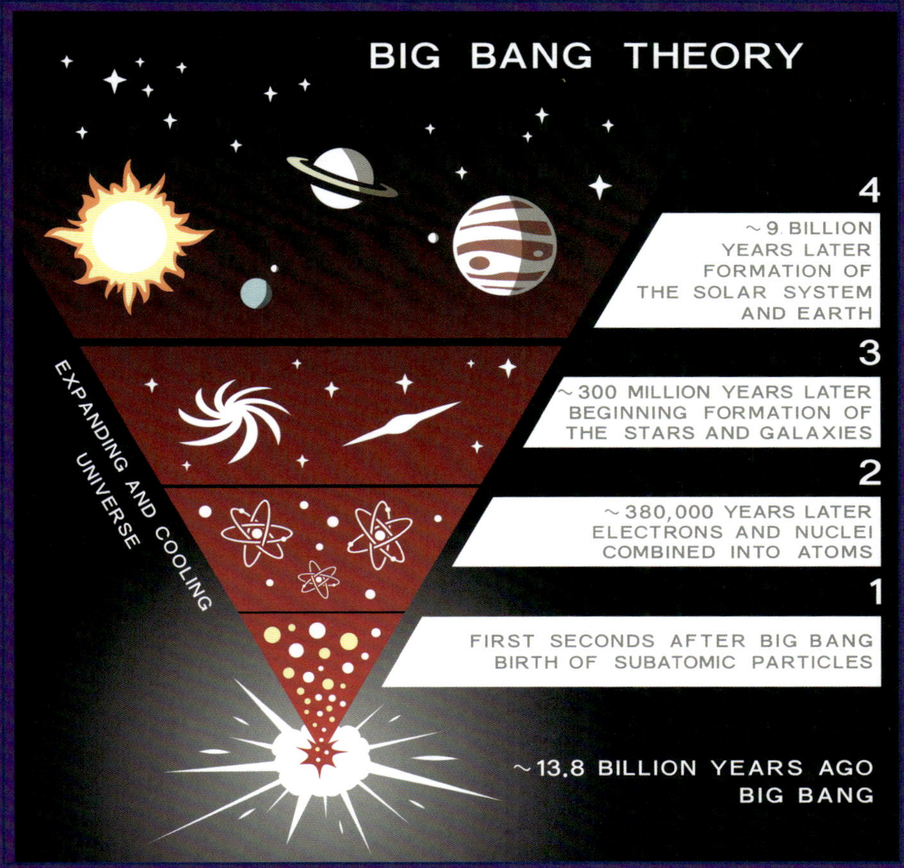

According to the big bang theory, rapid changes took place in the state of matter within a trillion-trillionth of a second. Scientists can observe the effects of the first explosion to this day.

The Cosmological Principle

Another essential component of the theory of the expanding universe had been introduced by Polish mathematician Nicolaus Copernicus (1473–1543). Copernicus not only put forth the unthinkable theory that the planets orbited the sun, but he also introduced the concept that the universe appears the same from any vantage point.

Specifically, this cosmological principle led to a view of the universe in which the universe is both homogeneous and isotropic when viewed at the largest of scales. Restated, the objects in the universe react to events in the same way. The objects in the universe are also the same in any direction you look. When you look up at the universe from Earth, you are seeing the same principle in action that you would see if you were to view the universe from another planet, or even when in a different galaxy.

The significance of the cosmological principle is that it is the basis for our understanding of the universe. Scientists work with the understanding that what they observe is not unique to observations from their planet. What they observe is a representation of what they would observe from any planet. Not that the objects they see are the same—those depend upon where they stand—but that the way the objects behave and the forces at work upon them are the same, no matter where that scientist stands.

The cosmological principle is also significant because it includes the requirement that there are no special exceptions to the theories used to understand the cosmos. If there is something that is discovered that seems to be specific to one place or set of circumstances, it is accepted that there is some aspect of that observation that is not yet understood. This rigor is essential to an overall understanding of the cosmos. It also leads to questions that must be solved through further observations and study.[3]

Hubble's Constant

Hubble's work proved that the universe was expanding and that Lemaître's theory was correct. His work also showed, through the use of the red shift phenomenon, that planets farther away from Earth appear to move away more quickly.

The next question was, just how fast was the universe expanding? The Hubble constant (H_0) is the figure that Hubble derived for use in this calculation. It was based upon the belief that the universe was formed about two billion years ago. His work was also based upon the use of standard candles, the Cepheid stars he had observed. It wasn't until the 1950s that astronomers discovered that Hubble had mistakenly used star clusters, rather than individual stars. Because of this, not all of his objects had an absolute luminosity, least of all one that differed with how far away they might be.

Hubble had no way to know that anything he had done might be flawed. Neither could he have known that the estimated age of the universe would change once new technology revealed it was even older than thought at the time. Despite the impact upon his constant that these discoveries would have, Hubble estimated the rate of expansion of the universe to be 310 miles (500 km) per second per megaparsec, or 99 miles (160 km) per second per million light-years.

The value of the Hubble constant remained a point of interest. Work done at the Mount Wilson and Palomar Observatories in the 1960s and 1970s resulted in the value being reduced until it reached 34 miles (55 km) per second per megaparsec. Other astronomers, most notably those in Sydney, Australia, calculated the value near 62 miles (100 km) per second per megaparsec. Part of this difference was due to differences in methodology. Still, an accurate and mutually accepted value for the H_0 was essential to the work remaining.[4]

Cosmic Radiation and the Expanding Universe

Scientists had proof through observation of the red-shift phenomenon that they universe was expanding. They had a value for that rate of expansion in the Hubble constant.

In 1948, American cosmologists Ralph Alpher (1921–2007) and Robert Herman (1914–1997) measured the temperature of microwave background radiation. Their work was appreciated but not applied to the big bang until the 1960s, when Soviet physicist Yakov

Robert Wilson (*left*) and Arno Penzias pose in front of the antenna that first picked up the cosmic microwave background (CMB), proof of the big bang.

Zel'dovich (1914–1987) examined their research. At the same time, American physicist Robert Dicke (1916–1997) became interested in the cosmic microwave background (CMB), the remaining light from the big bang.

The first to detect CMB were American physicist Arno Penzias (1933–) and American astronomer Robert Wilson (1936–). Working for Bell Labs in 1964, they were using a large antenna to map signals from the Milky Way, but they were troubled by a noise that came at them from all directions. They tried turning the antenna in different directions and taking drastic measures to get rid of the pigeons who roosted in the mechanism. Finally, rid of everything that could explain the disturbance, the noise remained.

At about this time, Dicke contacted them before starting his search for low-level radiation as support for the big bang. When he heard of their discovery, he is said to have turned to his team and said, "Well boys, we've been scooped." Both groups published their results, but it was Penzias and Wilson who were recognized for their proof of CMB. It was they who received the Nobel Prize for their work.[5]

The HST Key Project

Prior to the launch of the Hubble Space Telescope (HST) in 1990, scientists put together a list of the projects they wanted to do with the telescope. The Key Project to Measure the Hubble Constant was of great importance. A more accurate value for the constant would help to determine the size of the universe, as well as what might happen to the universe in the future.

With the launch of the Hubble Space Telescope into low Earth orbit in 1990, this proposed project became a reality. The

HST could make observations from outside Earth's atmosphere. It could also be used to explore objects at a far greater range than previously available. After making adjustments to align the different methodologies, a value for the H_0, given what was known about the age of the universe at the time of the research, was set at 44 miles (71 km) per second per megaparsec, plus or minus 7.[6]

Chapter Six

What's Next in Dark Energy?

From the moment Penzias and Wilson detected noise they could not explain with their antenna at Bell Labs, scientists wanted to know more. Was this the background radiation left over from the big bang? Was the CMB spread evenly throughout the universe? Was the universe truly isotropic and homogeneous, or were we in a unique position in the cosmos?

The Cosmic Background Explorer

In 1989, NASA launched the Cosmic Background Explorer (COBE), also known as Explorer 66. This satellite was used until 1993 to search for additional data about the cosmic microwave background. COBE was not a telescope like the Hubble Space Telescope. It was a satellite that could detect heat in the universe. This heat was in the form of radiation. If COBE detected CMB, it would confirm the findings of Penzias and Wilson. It would also supply evidence to support the big bang theory.

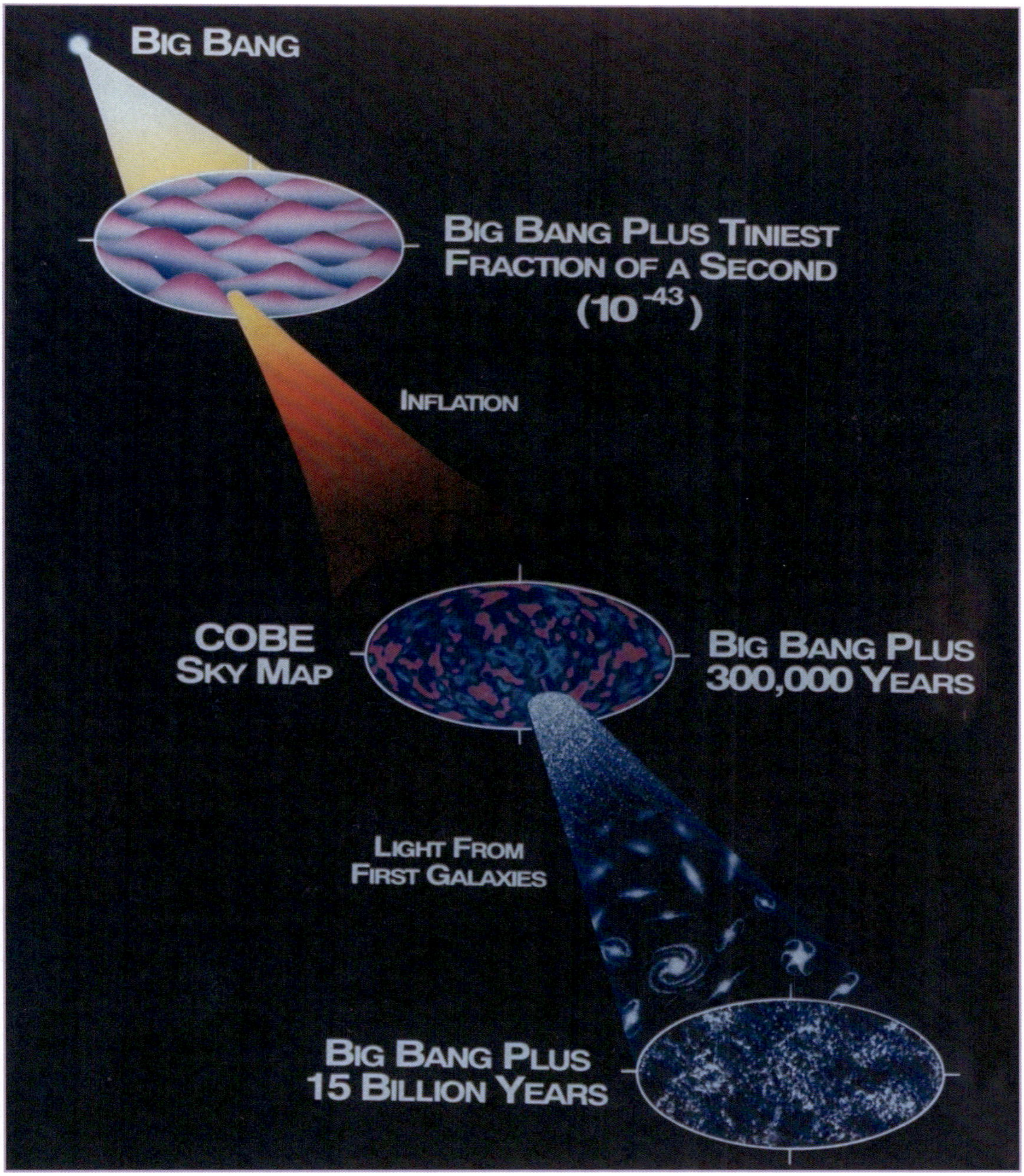

This image represents the expansion of the universe since the big bang. The COBE sky map provides a detailed image of the variations in temperature in CMB throughout the known universe.

CMB is an essential part of the big bang theory because theorists speculated that the light from the bang would be bright and hot. Over time, this light would travel away from the point where it originated. It would cool and its wavelength would show that, but it would still exist in the universe. If COBE did not find proof of CMB, then it might have been that Penzias and Wilson had discovered radiation from a different event. As it was, COBE did find evidence of CMB throughout the universe.

American cosmologists George Smoot III (1945–) and John Mather (1946–) received the Nobel Prize in 2006 for their work with COBE. As is often the case in science, their work opened the door to additional questions. As is also often the case in science, cosmologists' work suggested other things that should be found in the universe. In the same way that Einstein's theory of general relativity suggested a force represented by lambda, there were new observations that needed to be made to support other ideas.

Often, well-founded ideas are in place years before there is the technology to provide answers or proof. This was the case with the satellites and probes available for the search for CMB. Scientists knew that more powerful equipment would provide more precise answers and data.[1]

The Wilson Microwave Anisotropy Probe

NASA launched a second probe to study the universe in 2001. In operation until 2010, the Wilson Microwave Anisotropy Probe (WMAP) was designed to search for evidence that the universe was isotropic—that its composition was the same when viewed

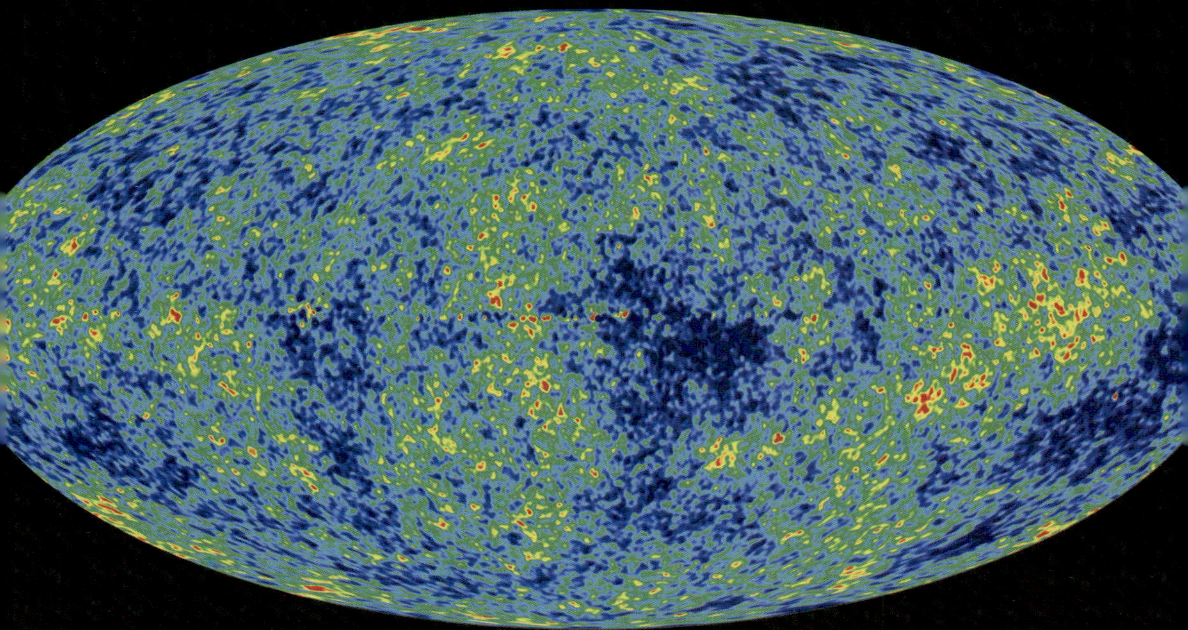

The WMAP provided images of significantly high resolution. This image reveals temperature fluctuations (red is hotter, blue is colder) dating back 13.7 billion years. It helps pinpoint when the first stars formed.

from all angles. This would support the big bang theory because after the big bang, the basic materials of the universe should have burst into space in a uniform manner. It would not have spewed material of different types in different directions because all matter was contained in that tiny point when the bang occurred.

The WMAP team made some significant findings. It was they who determined that the universe was 13.77 billion years old. They also calculated the curvature of space and discovered that ordinary atoms (baryons) make up about 4.6 percent of the universe. Dark matter, or matter not made up of atoms, accounted for 24 percent of the universe. They confirmed that the expected fluctuations in hot and cold spots throughout the universe formed a bell curve.

With confirmation of some points came discoveries that led to new questions. Not surprisingly to people who are familiar with quantum physics, the WMAP team found that the science that explained events at a large scale did not explain events at the smallest scale, and vice versa. They also found evidence of inflation of the universe—the expected finding that went with the big bang. The WMAP team also discovered that dark energy—the cosmological constant—makes up about 71 percent of the universe. It is this dark energy that is causing the rate of expansion in the universe to speed up. Finally, they discovered the existence of a cold spot.

WMAP is significant because of its level of precision. It also is significant because it provided proof that our universe has a very small amount of ordinary matter. Dark matter and dark energy far outweigh it, with dark energy having the greatest presence by far. Scientists from a variety of fields were excited by the findings. Brian Greene of Columbia University said, "In a sentence, the

THE LIMITATIONS OF TECHNOLOGY AND SCIENTIFIC KNOWLEDGE

It is not unusual for scientists to anticipate the existence of something that it is not yet possible to prove. Einstein's theory of special relativity needed a constant to take the place of what we now call dark energy. Hubble's constant was calculated at several rates until technology improved to the point that one calculation that satisfied all could be made. Bosons, very tiny particles from the big bang, were part of scenarios of the big bang for nearly fifty years until their existence could be proven through the use of equipment and science that came along.

The images from COBE, WMAP, and Planck show the differences that technology can make. Each of these images captured the same information. The difference in the images is due to improved precision in the equipment. The result of this ever more specific imaging is that new findings are made possible.

observations are spectacular, and the conclusions are stunning. WMAP data support the notion that galaxies are nothing but quantum mechanics writ large across the sky. To me, this is one of the marvels of the modern scientific age."[2]

The Planck Space Probe

The Planck space probe was launched by the European Space Agency (ESA) in 2009. In service until 2013, it provided the most precise view ever obtained of CMB in the universe. It provided the most detailed mapping of the universe ever attained. While providing proof for things scientists expected, Planck also provided a look at some things scientists hadn't anticipated. These were things that challenged our understanding of the birth of the universe.

Some of the work done by Planck supported work done by COBE and WMAP. For instance, while WMAP dated the birth of the universe to 13.77 billion years ago, Planck puts it at 13.82 billion years. There is a difference, to be sure, but the newest finding is in keeping with that of the WMAP. Planck also found that CMB is evenly distributed, as had been theorized.

However, Planck did not just bring us the greatest precision ever with the best map of CMB, it also showed that there were some anomalies. These differences should not exist if the current theory is correct. Planck not only observed the cold spots first observed by WMAP, it observed them with greater precision. This cold spot is not simply one of the many variations in temperature seen in the universe. It is a specific area that is larger than any other cold spot.[3]

What Is the Cold Spot?

Some scientists view the cold spot as the result of mathematical manipulation of the data—something that exists in the model but not in the actual universe. Some view the cold spot as potential proof for a multiverse. This would mean that our

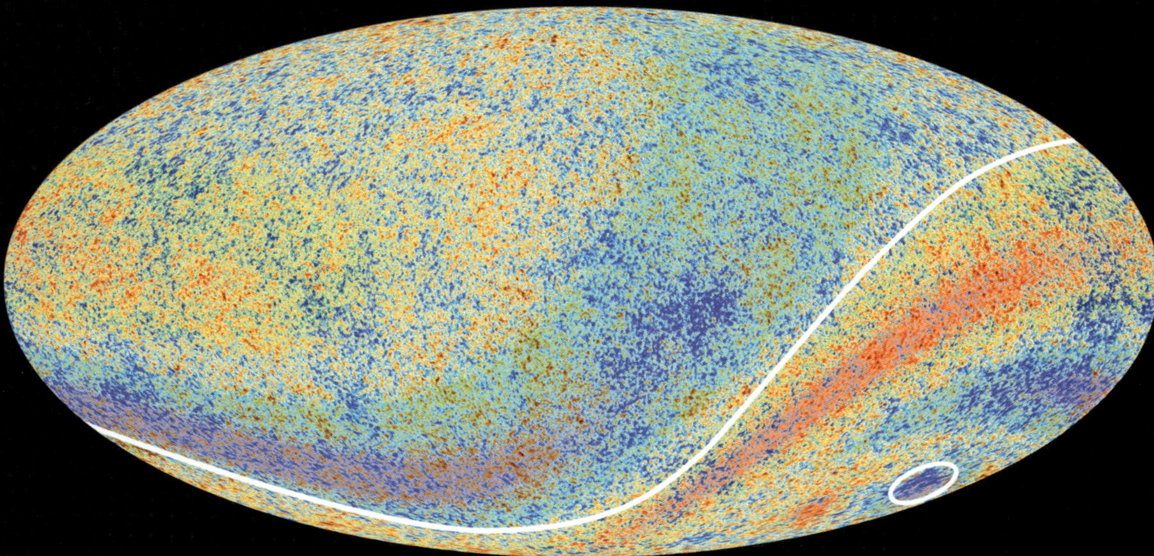

Anomalies throughout the universe are evident in this image captured by the Planck space probe. The average temperatures in the hemispheres of the sky (separated by the curved line) vary; it's slightly colder in the northern hemisphere (indicated by the blue shading), which doesn't align with scientists' prediction that the universe should be generally similar in any direction. There is also a cold spot, circled in the image, that is much larger than anticipated.

universe is not alone, that there are other universes, and perhaps two universes are touching at that spot. The existence of multiple universes opens the door to possibilities not envisioned until this point. To prove or disprove this will require more experimentation, as well as physics and tools we don't yet have.

In April of 2015, the Monthly Notices of the Royal Astronomical Society suggested the cold spot might be due to a supervoid in space. The research for this was based in part on a team at the University of Hawaii at Manoa led by István Szapudi. Szapudi's team wrote that such a supervoid would be evidence that the universe is expanding. The colder temperature would be caused by light traveling through the supervoid, losing energy as it did. Szapudi also said the supervoid would be "the largest individual structure ever identified by humanity."[4]

Some believe there are multiple universes, depicted in this illustration as bubbles. The orange webs are galaxy clusters. Perhaps a neighboring universe is touching ours, creating the cold spot recorded by WMAP.

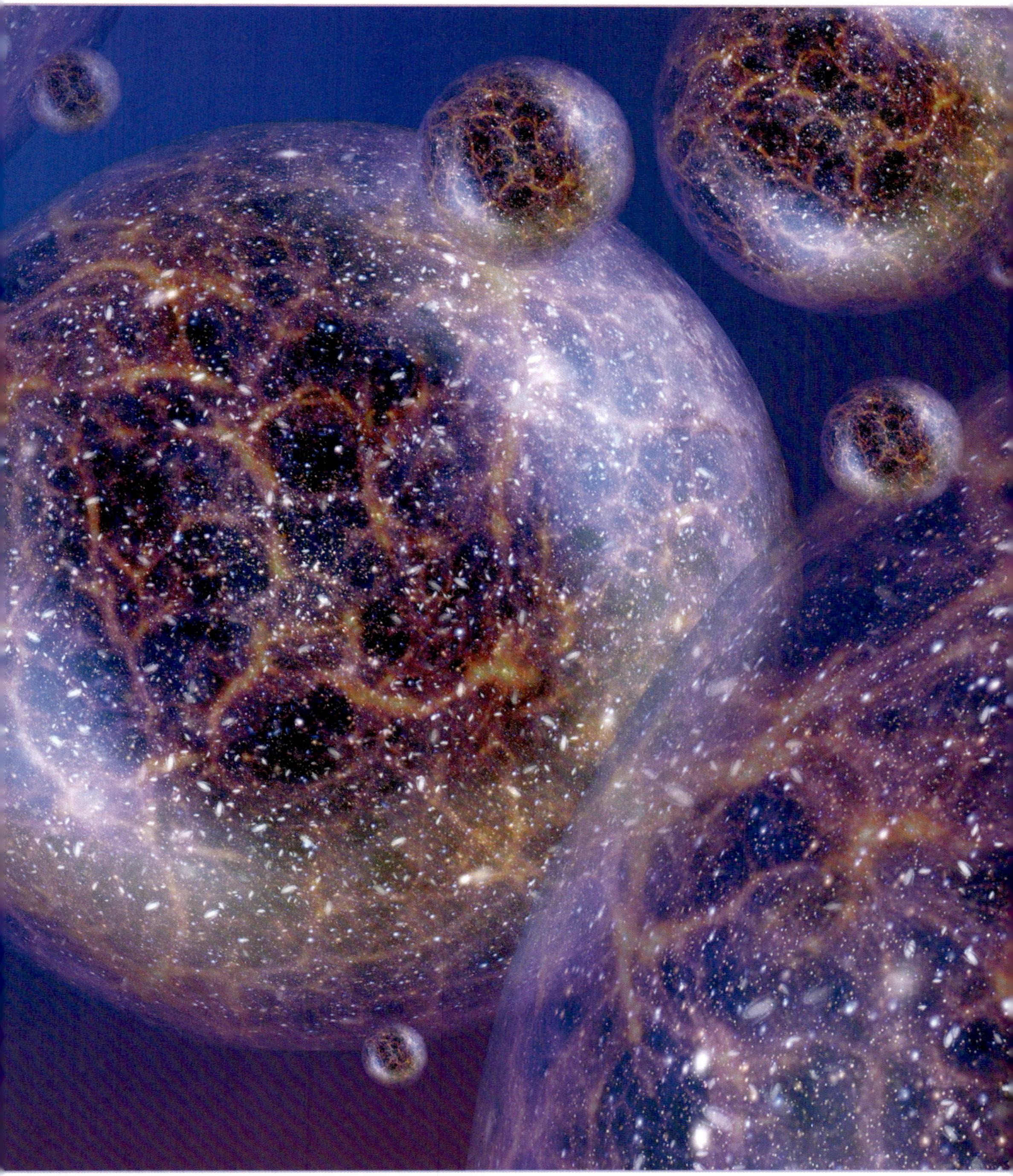

H_0 Versus Dark Energy

Once scientists determined that the universe was expanding at an increasing rate, they were at a loss. If their understanding of the origins of the universe was correct, everything that made up the universe had existed since that tiny speck of matter at the origin. How could more energy be added to this closed system? How could anything be speeding up? And, not insignificantly, what did this mean for Hubble's constant?

Over the decades since it was first calculated, H_0 has changed as new information is introduced. The constant is calculated as the rate of change of the distance between two points divided by the distance between those two points. As the change accelerates, the constant becomes smaller. How is it possible for the universe to be expanding while the Hubble constant is becoming smaller since the Hubble constant represents the rate of expansion?

One scenario points to the existence of dark energy. This phenomenon, which makes up more than 70 percent of the energy of the universe, could be the source of the energy used in the accelerated expansion. With our current level of technology, it is impossible for scientists to prove or disprove the role dark energy plays. Could it be that the original calculations were flawed? Could it be that there are forces in the universe we still have not identified? We must wait until we have the equipment and understanding required to make this determination.[5]

CHAPTER NOTES

Chapter 1.

Einstein, Noether, Hilbert, and Theoretical Cosmology

1. TheCosmicWeb, "Special Relativity vs General Relativity," YouTube, September 9, 2013, https://youtu.be/Eo3PV0HTI9U.
2. "Hermann Minkowski—The Man Who Posed the Problem," Concordia University, accessed May 2, 2018, http://www.spacetimesociety.org/conferences/2004/minkowski.html.
3. Matthew Smith, "The Thought Experiment," YouTube, May 1, 2012, https://youtu.be/dPFS01Hs35A.
4. Tech Insider, "Einstein's Blunder Explains One of the Greatest Scientific Revelations," YouTube, June 1, 2017, https://youtu.be/QcJTKKCtAil.
5. Natalie Angier, "The Mighty Mathematician You've Never Heard Of," *New York Times,* March 26, 2012, https://www.nytimes.com/2012/03/27/science/emmy-noether-the-most-significant-mathematician-youve-never-heard-of.html.
6. Luke Mastin, "20th Century Mathematics—Hilbert," Story of Mathematics, 2010, http://www.storyofmathematics.com/20th_hilbert.html.
7. "Alexander Friedmann (1888–1925)," Physics of the Universe, 2018, https://www.physicsoftheuniverse.com/scientists_friedmann.html.
8. "Edwin Powell Hubble—The Man Who Discovered the Cosmos," Hubble Space Telescope, accessed May 2, 2018, https://www.spacetelescope.org/about/history/the_man_behind_the_name/.

9. "Did Einstein Predict Dark Energy?" Hubblesite, accessed May 2, 2018, http://hubblesite.org/hubble_discoveries/dark_energy/de-did_einstein_predict.php.
10. "The Purpose of CTC," Stephen Hawking Centre for Theoretical Cosmology, accessed May 2, 2018, http://www.ctc.cam.ac.uk.

Chapter 2.
Hubble and the Expanding Universe

1. "Cepheids," NASA, accessed May 2, 2018, https://starchild.gsfc.nasa.gov/docs/StarChild/questions/cepheids.html.
2. Elizabeth Howell, "Henrietta Swan Leavitt: Discovered How to Measure Stellar Distances," Space.com, November 11, 2016, https://www.space.com/34708-henrietta-swan-leavitt-biography.html.
3. Clara Moskowitz, "Star That Changed the Universe Shines in Hubble Photo," Space.com, May 23, 2011, https://www.space.com/11761-historic-star-variable-hubble-telescope-photo-aas218.html.

Chapter 3.
Supernova Searches and Accelerated Expansion

1. Paul Preuss, "The Evolving Search for the Nature of Dark Energy," Supernova Cosmology Project, October 27, 2009, http://newscenter.lbl.gov/2009/10/27/evolving-dark-energy.
2. Brian Schmidt, "The High-Z SN Search," Harvard-Smithsonian Center for Astrophysics, accessed May 3, 2018, https://www.cfa.harvard.edu/supernova/home.html.

3. Andrea Thompson, "What Is a Supernova?" Space.com, February 8, 2018, https://www.space.com/6638-supernova.html.
4. Ibid.
5. Schmidt.
6. "The Nobel Prize in Physics 2011," Nobelprize.org, accessed May 3, 2018, https://www.nobelprize.org/nobel_prizes/physics/laureates/2011.
7. P. Nugent and SNAP Collaboration, "SNAP: Supernova/Acceleration Probe an Experiment to Measure the Properties of the Accelerating Universe," *AIP Conference Proceedings* 540, no. 1 (March 2001), https://aip.scitation.org/doi/10.1063/1.1328889.

Chapter 4.
The Cosmological Constant

1. Cormac O'Raifeartaigh, Michael O'Keeffe, Werner Nahm, and Simon Mitton, "One Hundred Years of the Cosmological Constant: From 'Superfluous Stunt' to Dark Energy," *European Physical Journal H* (December 25, 2017), pp. 17, 31, https://arxiv.org/pdf/1711.06890.pdf.
2. "Did Einstein Predict Dark Energy?" Hubblesite, accessed May 2, 2018, http://hubblesite.org/hubble_discoveries/dark_energy/de-did_einstein_predict.php.

Chapter 5.
Evidence of the Big Bang

1. "Profile: George Lemaître, Father of the Big Bang," American Museum of Natural History, accessed May 3,

2018, from *Cosmic Horizons: Astronomy at the Cutting Edge,* ed. Steven Soter and Neil deGrasse Tyson (New York, NY: New Press, 2000), https://www.amnh.org/explore/resource-collections/cosmic-horizons/profile-georges-lemaitre-father-of-the-big-bang.

2. "Origins of the Universe: What Is the Big Bang Theory," National Geographic Science, accessed May 3, 2018, https://www.nationalgeographic.com/science/space/universe/origins-of-the-universe.
3. "Nicolaus Copernicus," Biography, updated March 15, 2018, https://www.biography.com/people/nicolaus-copernicus-9256984.
4. John P. Huchra, "The Hubble Constant," Harvard-Smithsonian Center for Astrophysics, 2008, https://www.cfa.harvard.edu/~dfabricant/huchra/hubble/.
5. Alaina G. Levine, "Arno Penzias and Robert Wilson," APS Physics, 2009, https://www.aps.org/programs/outreach/history/historicsites/penziaswilson.cfm.
6. "A Decade of Discovery 1990–2000," Hubblesite, accessed May 3, 2018, http://hubble.stsci.edu/hubble_discoveries/10th/vault/timeline.shtml.

Chapter 6.
What's Next in Dark Energy?

1. "COBE," NASA Science Beta, accessed May 22, 2016, https://science.nasa.gov/missions/cobe.
2. "Wilkinson Microwave Anisotropy Probe," NASA, updated December 22, 2017, https://map.gsfc.nasa.gov.

3. "Planck," NASA Jet Propulsion Laboratory, accessed May 3, 2018, http://planck.caltech.edu.
4. Heather Gross, "What Created the Universe's Cold Spot?" *Air & Space,* June 2015, https://www.airspacemag.com/space/what-created-universes-cold-spot-180955343.
5. Jeremy Mould, "The Hubble Constant and Dark Energy," *Publications of the Astronomical Society of the Pacific* 123 (August 22, 2011), IOPScience, accessed May 3, 2018, http://iopscience.iop.org/article/10.1086/661992/meta.

GLOSSARY

big bang theory The theory that states that the universe originated from an explosion of a single point 13.8 billion years ago.

Cepheid A star that grows brighter and darker on a regular cycle.

cosmic microwave background (CMB) The remaining light from the big bang.

cosmological constant Also known as lambda (Λ), it is the fudge factor used by Einstein to balance his equations for his theory of general relativity.

cosmological principle The scientific principle that states that the universe is homogeneous and isotropic.

cosmology The distinct field of science that is concerned with the start and behavior of the universe.

dark energy A theoretical force that acts against gravity to cause the universe to expand at an accelerating rate.

dark matter The material in the universe that does not contain atoms, that is not visible, and that we are aware of because of its actions upon us.

finite Limited in quantity.

homogeneous Having all parts made of the same material.

infinite Unlimited in quantity.

isotropic Being the same in every direction.

matter Anything that contains mass.

multiverse The idea that there are an infinite number of universes in existence.

red shift The result of a body moving away from Earth very quickly. The farther away the star is, the longer the light waves have to be to reach Earth. This longer wave appears red to us on the spectrum of observable light.

supernova The explosive death of a star.

supervoid A massive region in space that has very few or no galaxies.

symmetry A concept in physics that states the properties of a physical system stay the same after going through a transformation.

universal laws of conservation Several scientific principles that state that a certain measurable property of an isolated physical system does not change over time.

FURTHER READING

Books

Blumenthal, Kelly. *Cosmic Inflation Explained.* New York, NY: Enslow Publishing, 2018.

Lew, Kristi. *Dark Matter Explained.* New York, NY: Enslow Publishing, 2019.

Saxena, Shalini, ed. *Cosmology: Understanding the Evolution of the Universe*. New York, NY: Britannica Educational Publishing, 2014.

Sparrow, Giles, Judith John, and Chris McNab. *How the Universe Works.* New York, NY: Cavendish Square, 2016.

Wells, Robison E. *Dark Energy.* New York, NY: Harper Teen, 2016.

Websites

Euclid Consortium

www.euclid-ec.org

Learn about Euclid, the European Space Agency's mission to discover why the universe is expanding at an accelerating rate and what is responsible for this phenomenon.

NASA

www.nasa.gov/subject/6891/dark-energy-and-dark-matter

Read about the latest research and discoveries on dark energy and dark matter.

Space.com

www.space.com

Dive deeper into space exploration, innovation, and astronomy news.

INDEX